PHYCHO

Ashlesha Prabhu

This is a work of fiction. Names, characters, places, and incidents either are the product of the author's imagination or are used fictitiously. Any resemblance to actual persons, living or dead, events, or locales is entirely coincidental.

Text Copyright © Ashlesha Prabhu, 2022
Illustrations Copyright © Pushpendu Mandal, 2022

All rights reserved. No part of this book may be reproduced or used in any manner without written permission of the copyright owner except for the use of quotations in a book review. Requests for permission should be addressed to ashleshaprabhu@gmail.com

First paperback edition March 2022

Book design by Ashlesha Prabhu & Pushpendu Mandal

ISBN: 979-8-4290-1242-1 (paperback)

I DEDICATE THIS BOOK TO MY GRANDFATHER, WHO, TILL HIS LAST BREATH, BELIEVED IN THE MAGIC OF LIFE. HE ACKNOWLEDGED MY FIGHTING SPIRIT; I ADMIRED HIS.

"YOU DON'T HAVE TO FIGURE THINGS OUT. THINGS WILL GET FIGURED OUT FOR YOU. LIFE IS ABOUT *YOU*."

<3

Author's note

Phycho was born about a year and a half ago when the world had come to a halt. So had my career. Indeed, the dark times had arrived. My love for education had left me with a master's degree in psychology and a bachelor's degree in physics. However, only a few close ones knew about my talent in writing and my pursuit of acting. I am someone who always has refused to fit in the box. I have always wanted it all. I still do. That is what scientists do; They experiment. Not knowing what to do with the inadequate knowledge that I thought I had, I ate my mother's head ever so often. Certain events that followed made my life take an ugly turn. Even as a therapist, it took me time to realize I had been suffering an existential crisis. Questioning the meaning of life, the purpose of living, survival, and cruelty had become my routine. It later also became my area of research.

As a psychologist trained in children's psychology, I always come across brilliant thoughts born out of these curious minds. And then, one day, it hit me. I realized that one effective way to target cruelty in the world is to help raise a generation of kind, determined and hopeful children. Every coming generation is the biggest hope for all the living. It is our teenage years when the whole world starts to seem like an enemy. Therefore, I believe that if I could, in my own possible way, instill values, cultivate empathy and restore faith in the future of tomorrow, I would be looking at an even kinder, wiser, and more hopeful world.

As a child, too, I had a vivid imagination. The only people who heard my bizarre stories then were my parents and sister. Now, I bore my friends and the world. One of the by-products of such imagination is coming up with theories. As a student of both science and arts, I see a lot of similarities and contrasting ideas that these fields have to offer to us. In this book, I have chosen to address the fundamental problems which change the course of the development of teenagers by equipping them with emotional tools to battle the storm within. It is where the wisdom from these theoretical ideas comes in handy and kindles their academic thirst. It is my sincerest attempt to be helpful to anyone who reads it in a time of need or revisits it during one.

Phycho is not only purpose-driven but also a good read for all. If you never enjoyed learning basic physics concepts in school, this book has it covered. Additionally, it has introduced a few ideas from psychology that are taught at the pre-college level. Why put such advanced concepts in a book primarily for pre-teens and teens? Firstly, I believe psychology is a subject that needs to enter middle school level education as per the changing times. Secondly, it is again my belief that learning physics should never feel difficult. It entirely depends on who your teacher is. Finally, what is the use of a complex academic syllabus if we are unaware of its beautiful real-life applications? Turns out, there are many! Those strike us in mysterious ways later in life. So, I have decided to share my learnings that find hilarious analogies to academic studies. Whether you are a child, parent, book fanatic, book critic, hate science, love science, enjoy fiction, enjoy non-fiction, indulge in self-help content, or dig humor, *Phycho* welcomes and entertains all.

A child who hates reading, too, finds it impossible not to skim through the pages that have drawings. I used to be one of those. *Phycho* has stimulating illustrations to keep you engaged. It has educational information, well-researched ideas, solutions that work, a professional touch, and a fictional character. I have used my child-like alter-ego Phycho to write the book so that the young readers can relate. Amidst all the existing teenage comic characters, I have tried to introduce this new fictional character purely to cater to the need of these buzzing minds of being entertained. I hope to bring smiles to the faces and spread gratitude in the hearts of those whose hands acquire this little treasure of mine. Much love to all the children.

Contents

Prologue	1
Life	3
Relationships	13
Time	25
Thoughts	35
Perspective	45
Desire	55
Identity	65
Death	75
Epilogue	85
Acknowledgments	89

Prologue

Dear reader,

Introductions are meant to be lengthy, thorough, blah blah, and therefore, mostly *boring*. Well, who said I was appealing? My crush in school certainly doesn't think so. Anyway, I am writing this journal/diary/magazine or call it whatever you like because *I have a purpose!*

I may not be interesting, but I am definitely interested in you. All of you! I am interested in helping you discover yourself at an early age. I guess self-help is the new cool now. Also, I believe we all are similar when it comes down to our core. The world says we are the future. It is better to know a few things about what we are expected to do on our journey to becoming adults. Adults, if you are reading this, I welcome you too. Have a look at my way of understanding the world.

I wonder about life, events, people, dialogues, and in general, everything. Then, I use my understanding of physics and psychology to truly decipher the meaning of my experiences. Slowly, but every time, I find a way to cope with my existential issues. So, in this journal, I am writing about some simplified stuff about our complex existence and pieces of life advice.

On a parallel track, you might start to feel how this book would have started out as a personal diary. It did! When I start talking about ideas, I usually babble about myself and my issues. You will realize what I mean the further you read.

See, I started the lengthy and boring. To cut it short, just remember to picture me in your head as you move ahead on the chapters. Keep updating that picture until the end, and maybe you will start to see how similar you and I are. Wink!

xoxo,
(Adults, we use this to end our letters nowadays;
it's trendy!)
Phycho

Life

"C'mon, that's life!"

Have you heard of this sentence before? It is as if people want you to forget everything that just went wrong, all that turned your life upside down, or everything that made you start all over again. It is as if everything unfair and unbelievable was just made okay by that one simple sentence. Life then starts sounding like this evil, unfathomable twist of faith. So, what must you do when life is unfair?

Do not worry! I have got you now. Hello! I am Phycho! My father is a physicist. He analyses life using **scientific reasoning**. However, my mother is a psychologist. She understands life through a **humanistic approach**.

Although they both follow such individualistic, unique but interesting beliefs, sometimes you need to mix things up. Therefore, I have decided to call myself Phycho because I blend these ideas to derive unique interpretations of my experiences. I believe I am on a mission to share my understanding with you. It could help you reflect upon your own experiences.

If you ask me, there is not a single issue that we encounter in life which cannot be explained with the help of these two extensively researched fields of study. Unbelievable? Well, *C'est la vie*! (It is weird how the exact phrase sounds acceptable or, one may say, even charming in French.) So, what wisdom do our study buddies, physics and psychology, lay on us regarding the value one must give life?

I have not actually studied enough physics, but dad talks about the principles, theories, and his research all the time.

While learning Physics, we understand that human is made of tiny atoms and molecules. We have even discovered subatomic particles. On a microscopic level, the living cells in our bodies have a composition of these molecules. There also exists stuff like enzymes, hormones, and whatnot. All of which helps us survive. It is as if all these tiny *people* (Let's call them people because it sounds cuter) make us who we are and prepare us for who we want to be. We are metaphorically their whole universe.

On a macroscopic level, each individual on this planet is, in fact, super tiny in comparison to the size of the actual universe. There also might be more like us somewhere out there that we are not even aware of yet. So, considering the whole universe, **we are as good as we don't even exist**! Then why do we even pride ourselves so much? Who do we think we are??

Psychology will tell you who you really are! It is the study of human behavior, the human mind, our egos, our issues, our fears...blah blah blah. In short, it is **entirely about us**. Here, we are supreme. If one cannot care for oneself, how will one care for others? If you can't love yourself, how will you love others? It doesn't matter how big the universe is; we do exist! Our problems exist! As long as we live and even after we have lived, we exist on various levels. That's who we are!

These are two contrasting yet significant views about the value and size of our lives. However, now these two opposites will provide a solution on what to do when life is unfair to you!

1) **Don't take yourself too seriously**: We all do this because we value ourselves so much. It is *without a doubt* a healthy thing to do. You must give value to yourself; however, not so much that now the whole world revolves around you. Yes, you made a mistake. And yes, your plans got crushed. Oh! Yes, you might now feel like a loser. But that is all in the present moment. You were someone else in the past which is enough proof that you will be someone else in the future. Someone older and wiser. Someone who will definitely need this experience as a reminder at some point. Also, mom never forgets to mention that

people's memory of your life is short! So, no one will remember whatever happened after some time. If still they or you do, realize you are just not that important in this grand scheme of life. So, c'mon you are not the universe; you are just a part of it. Some outcomes are totally out of your control. Snap out of it! You *have* to.

2) Now that you have not taken yourself too seriously, **take yourself just a little too seriously**: Wait! What does that mean? Why did I say that? When you realize how tiny you are, it is time to reconsider yourself. Separate yourself from the problem. Remember, you and your problem are two different entities. Whatever your problem is right now, it still doesn't define who you are.

Were you going to pack your problem in a bag and carry it around the world wrapped around you forever? That sounds like my dad, who tries to fit things like tomatoes and potatoes in his pants pockets. Whenever he forgets to carry a bag to the supermarket, he imagines how mom would scold him later for spending extra money on the grocery bag the place provides. He has no other choice but to stuff the items in his garments and embarrass the co-shoppers like me. Anyway, is there even a market where you could find such a large grocery bag? Would you be able to fit in it the problem, the big ego associated with it, and the hurt that sometimes feels like a huge rock crushing your heart? Instead, clear those thoughts and use that space to store some new memories and lessons. If you have already managed to find that bag, use that space for souvenirs from those trips you will be taking. Or for your leftover food which was just too tasty to be left on the plate there. Or for the random number of photos of those once lived moments (some filled with crazy laughter that made you pee your pants; Maybe then, consider stuffing that bag with a diaper too?)

Remember, without you, those happy times were not possible. Remember how much you mean to someone. Remember how much you mean to yourself!

3) **Always ask yourself this when you are done doing the steps above**: Life may be unfair but are *you*? Don't be unfair to yourself because that's not you.

In short, you will look at yourself differently for a period when things go wrong. It takes time to grieve, calm down, breathe, reenergize and put things in perspective; Even our tiny people need time to replenish their used-up resources.

Nevertheless, you also need to decide on what amount of this reboot period is enough. In short, you need to fix how much time you will dedicate to your reboot period. We have a gift: our thoughts, our actions, our beliefs, our experiences, all that we are right now, and all that we are becoming. *This gift* is not something your life or the universe bestowed upon you. It is what you fought for, created yourself, and gifted yourself from the bits and pieces that *you* gathered from life and the universe.

All these words... Easily said than done! But wait, that means you have at least tried doing those at some point. Now, who cares if you fail, you took the first step! "You will get there." I'm so sorry! Didn't mean to leave you with another hook sentence. But hey, Phycho knows what to do when someone says that too. Wink!

Relationships

"The name is Bond. James Bond."

I am sure you are familiar with this famous movie dialogue. Bond's charisma helped him literally bond with his audience, and this introduction piece stayed with them even after the movie. Well, *the Bond* is central to the James Bond movie and for our next topic of discussion, relationships!

You must have heard many times that *human is social animal.* It is true because we have needs for love and belonging. Even so, people disappoint us, betray us, and the connection that once was unbreakable becomes weaker and weaker until it is non-existent. As a result, you might have found yourself uttering the following words. *"I cannot trust anyone ever again." "I don't have a single friend in this whole wide world." "Life will never be the same now that this is over."* You are absolutely correct with that last sentence. Life will never be the same. Why should life be the same anyway? Life is nothing but constantly changing; it was never meant to be the same because you are designed to evolve into something better. Don't you agree?

Well, back to bonds. What does physics say? Oh! Physics would definitely say, "Don't you get me started on this!" A pal of Physics - Chemistry - keeps repeating stuff about how atoms come together to form a bond resulting in new molecules. Molecules then further bond to form compounds.

Consider this: A carbon atom in the air turns into a muscle in your arms. How?? Another dear pal of Physics, Biology, will help you understand that. You see, a carbon atom from air bonds with oxygen atoms and transforms into carbon dioxide gas. Plants inhale this gas to produce food. These plants are, in turn, food for animals like cows. In the animal body, the carbon atom is essential to form protein. When humans eat animals or consume animal byproducts like milk, they obtain protein. Protein is required for the wear and tear of muscle tissues. A human then sends the carbon atom back into the air (for example, after exhaling carbon dioxide gas). Basically, a carbon atom evolves. It forms new bonds, breaks bonds, moves on, and serves its purpose. This is a **life cycle of a carbon atom**. Could this be similar to our life cycle too?

In physics, there are specific forces that help bring these molecules together. In people, the forces bringing them together are love, care, kindness, faith, and similar. The forces driving them apart are anger, jealousy, suspicions, and whatnot. Tremendous energy is also required for breaking these bonds. Well, tell me about the energy it takes to break and mend relationships! It takes so much patience, time, and effort to firstly build and later to gain back that lost trust. We are just like those tiny atoms and molecules that make us. I like to think that if a human were to be a world, then each living cell is its person.

All this talk about science... I am already missing its better half, Psychology.

Psychology also has some valuable information about relationships. Relationships are very crucial but not ultimate. Every relationship teaches you. It facilitates your emotional growth and development, offers valuable lessons, and prepares you for further endeavors in your future relationships.

A psychological therapy called **interpersonal therapy** is curated to address interpersonal problems. Mom explained the **model of interpersonal counseling** given by George Levinger to me the other day. He believed that the natural development of a relationship is in five stages: 1) acquaintance, 2) buildup, 3) continuation, 4) deterioration, and 5) termination (separation or death). Not all relationships go beyond the acquaintance. Likewise, not all deteriorate and terminate. Some friends are our 'Hi... How are you?... Bye!' - type friends whereas some are so close that you can never even think of separation. We all choose what kind of relationship we want with a particular person. However, it is a decision made by both individuals. Then there are times when the unimaginable happens. You end up being best friends with people you never before spoke a word to. Also, your best friends leave you to never return.

Physics and Psychology, thus, offered us the basic rule of nature: Bonds adapt over time. If we are to find an analogy, a person like an atom also serves different roles with changing times in each relationship. Therefore, one evolves. *You have the power to make a choice, but you cannot choose the exact consequences of those choices.* You, but then, also get to choose the method of how you will deal with those consequences.

So, how should we handle our relationships?? Here is the key: **Form relationships and break relationships and then form relationships again**! Something you already knew? **But what exactly should you do during each stage of this cycle**? When you want to start a relationship, be ready, show acceptance, be open-minded, be hopeful and put enough effort to establish an *appropriate* bond. When in a relationship, be fully dedicated to your relationship. You have to *truly live* a relationship like a carbon atom *serves its purpose*. So, explore, engage and evolve together. When the relationship blossoms, you will experience an assortment of emotions associated with it. Now, do not hold back, ditch the doubts, and embrace the newness with an open heart.

Let us just say, you and your friend are not talking to each other nowadays. Have you reached a point in your relationship where the person you once loved is now a person you do not love at all? Do you have anything left in you to give? If yes, please do so. You must try everything possible from your end before giving up. If not, then just stop, and start the process of acceptance that things have changed. In short, **quit the dependence**!

Understand that relationships cannot be broken in the true sense since they have already been registered in space and time in the form of memories. **Any relationship will always remain a small part of you and will continue to exist in one way or another.** But, **your emotional dependence on those relationships can be controlled by your own will**.

The good memories that you have of your friend or parent or that person will help soften your anger and pain. A day will come when you two will see past your differences. One day, you will notice a place reopening in each other's hearts. Go for it! Put in enough effort again. Retrieving the memory of my own experience, I believe I have still not forgotten what had happened because that something taught me a valuable lesson. However, I have learned to forgive. One action which feels extremely difficult but should not be that way is learning to apologize for your doing. People take time to do it, but you can choose to forgive them despite their slow speed; They don't even have to know. Let them reach there when they are ready. Meanwhile, you enjoy the view at the peace station!

When you do form a newer, more appropriate bond with the same person, you might find yourself thinking that you will never go back to what you two were before. And why would you? That will just remove the possibility of the creation of something new. Maybe you both just interacted/combined in the wrong way like innumerable atoms do every second, and that resulted in a weak bond. However, if you keep assuming that this will happen again, you have now just paved a path for suspicions and doubts.

You should aim to create a more **appropriate bond** this time; hopefully, a stronger but suitable one. You might notice that the chemistry between a pair may not be correct if lovers, but the chemistry is mind-blowing if friends. Look at Physics and its friends! Physics, chemistry, and biology are thoroughly interconnected; they have successfully formed these admirable bonds and are ready to help us with examples.

What happens in the next stage? When a relationship ends, the pain will fade with time. *It is perfectly alright* if you still wish to have that person in your life, even in the tiniest possible way, knowing well that nothing you do can make that happen. Any progress in that direction might be beyond your capacity for now. You need enough time and space.

Unless someone was a threat to your survival in some way, there are second chances. If someone is a threat, never give it a second thought; forgive and put an urgent full stop to that relationship. Save those efforts! Nature teaches us survival first. *Only* if you survive, you may evolve.

Once you *evolve*, you will see your past relationships, mistakes, and experiences in a totally fresh new perspective that you could have never imagined before. You will start noticing your own progress when you implement the lessons you learned, hold on to hope, and are ready to give again. You will then look back onto your old version and might even let out a chuckle the way we do when we look at our fascinating ancestors. You know who. *Thank you, evolution!*

In a relationship rollercoaster, there is a rise after every fall. I mean, there are always new people filled with stupendous energy from the universe like you are. Go! Meet them. Form new bonds. Live your life cycle. Do not forget to give them a little introduction like James Bond did! Or like I, your new pal Phycho, did earlier. Wink, wink!

24

Time

"I wanna go back in time and change what happened."

Each of us has thought or said this at least once in our lives. Then, some people seriously went a step further to work on a time machine. Yes, the research is going on, and we all might discover something very bizarre about time after all. Nevertheless, at present, for us, the only perceivable time is that of the present and the future. Past has already occurred and is now stored in our heads. Interestingly, by that logic, memory can be called *the time machine of our brain to travel the past.*

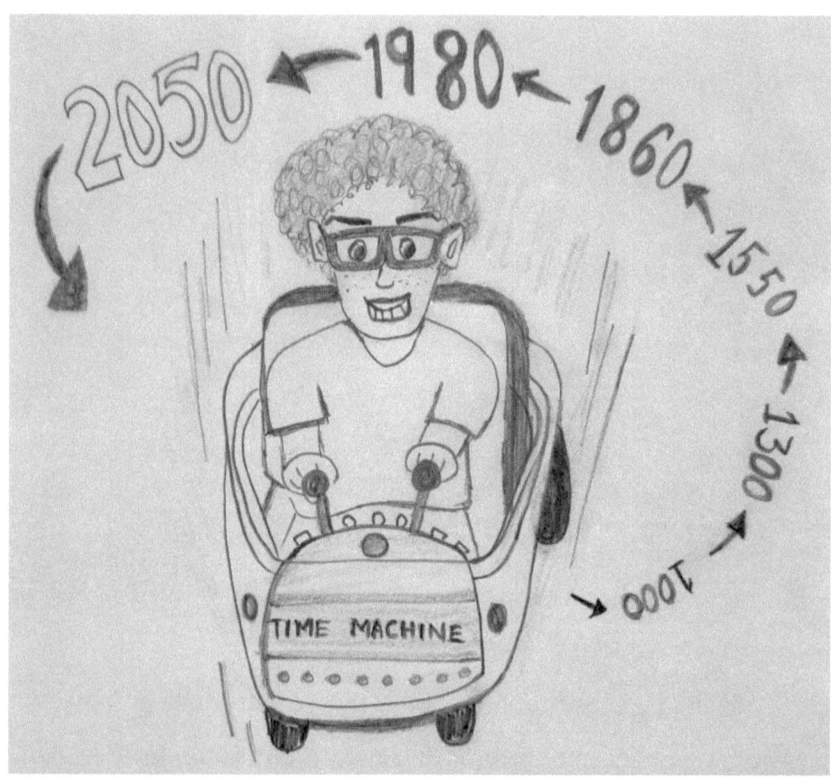

Many such interesting ideas and theories bring my attention to my fabulous friend, Physics. The famous string theory conceptualizes time as a dimension similar to height, length, and width. Interesting? One of the greatest minds ever, Einstein, once wrote: The distinction between past, present, and future is only a stubbornly persistent illusion.

So, we are just moving forward in the time dimension. For example, imagine watching a movie. Now, when you forward it, different incidents have occurred, people have gone through some changes from the original *reference point*. The part you are watching now is *present* but was the *future* for you before. If you reversed that movie, you would see that something totally different kept happening before the reference point. The *past* was also the *present* at some point.

Physics has also made it possible for us to measure time. We already know that because of our fancy watches. **It also helps us understand time in such depth that we can try and slow down the time or fasten it**. Yes! You heard me!

Again, you must have seen this concept in movies where a person travels across space through a wormhole, which is a theoretical tunnel connecting two distant regions in space-time. On his return, he looks like he has aged only a little, whereas everyone else on earth has aged noticeably. In fact, there are new generations of kids and grandkids ready to meet and greet this person now. The movie is called *Interstellar,* by the way. Considering reality too, if we travel in an insanely fast airplane, time goes *slower* for us, and we have thus aged a little slower than others. However, various other conditions are involved, and this measured time difference is smaller than a millisecond, even for the crazy fast airplanes.

Try searching these cool physics experiments on the internet for the exact numbers. The time difference is so less that it is considered insignificant. But hey! That is how humans can slow time! We have accomplished so much.

We have learned how the same time passes relatively different for each of us. Doesn't it already? A good time for somebody can be a difficult time for someone else. Whenever I think of this, my mind pulls up only that one memory. My basketball teammates always enjoy their extra practice sessions where my attention is solely dedicated to my screaming leg muscles. Or like when yesterday, I requested the coach for some more swim time (*easy peezy* for me) and annoyed the entire team. Apparently, they got mad at me, thinking the precious thirty minutes of their lives were wasted. (Thank me later, leg muscles.)

We all know that even though we cannot totally control time yet, in a way, we can, and we already do it; We make timetables for time management. Also, people out there, who are highly emotionally dependent on their schedules, now chill. Someday a ride in a super-fast plane might just save you some milliseconds and make up for that lost time. There is still hope for you.

Psychology, my other pal, works on the magical principle of how we have the **power of traveling through time using our imagination and memories**. We imagine our plans and the consequences of our decisions. It grants us a peek into the future that possibly can be true *tomorrow*, provided your imagination is relevant and reliable. That is what makes companies plan their future goals, teachers plan their school calendars, and people schedule appointments. They envision a possible future. Imagination triggers your creativity, provides insights, facilitates problem-solving, makes inventions and discoveries happen, and whatnot.

Now, if you imagine getting magical powers in the future, let us agree that you are being unrealistic. Nevertheless, it is so pleasing to just be able to imagine something which seems unreal right now: A future that may or may not exist. It gives you amusement, hope, motivation and does these weird wonders in your brain, filling up your whole body with so much excitement.

What about the *past*? Quite Simple. Our memories are the key to retrieving the events from the past so that we can align them in our heads exactly how they had occurred. It is how we can remember important information about what was said, heard, viewed, or felt. This information later helps us in our professional lives when writing books, making paintings or creating movies. An entire crime scene can be retraced in mind to obtain insight into the missing clue. Our memory helps us remember phone numbers, faces, places, events...phew! Our brain is capable of so much.

Although all this information is profoundly illuminating, let us move on to the life lesson that I would love to share with you. It came from a first-hand experience.

I had a terrible argument with my best friend because she lied to me. So, when I told her to never show me her face ever again, she accepted. I was furious. I wished to go back in time and change everything that had happened. I wanted that if I were to go back in time, I would have never made her my friend in the first place. That wish slowly changed to if I were to go back in time, I would have done things differently to change her perspective. I believed that then she would have never felt the need to lie. I had started developing empathy and reminiscing my love for her because our wonderful memories kept coming back to me.

Wait! So, in fact, **I *did* go back in time**, but only through my memories. Eventually, I quit wishing to travel back in time because I no longer knew what exactly I wanted to work out differently. In due time, a series of events after the fight helped me in more than one way. I learned a few things about myself, understood how situations work, and felt I had grown as a person. The pain also started fading after our joyful memories flooded my brain. I finally acknowledged what I was supposed to realize and said it out.

"I **am the one still carrying the pain in my present. And if *I* do not stop dwelling in the past, *I* will be dragging that pain in my future."**

I had traveled to the *past* for a long duration now. It was time to keep that pain aside and book my tickets for the trip to *the future* via my strong imagination of a promising tomorrow. So, whenever you wish to travel back in time and change things,

1) **Do that**, using the power of your memories and imagination. This will help you imagine a thousand parallel timelines that make things perfect until you realize that this is consuming your precious time and energy from the present moment. As soon as you recognize that whatever happened triggered a series of some beautiful events that otherwise would not

exist, you will *finally* let it go. (Think hard! You will find at least one, even in the worst cases.)

2) **Let it go**. Be grateful for memories and lessons that time delivered. Your future awaits you.

It is marvelous how we can write/play our own story in our head like a movie. Then, go ahead, live life, and update/edit that movie with some fascinating plot twists. If you ask me now if I were to go back in time and change something, I shall not change a thing because today, I get to be your pal, Phycho. I absolutely love it.

Thoughts

"If I don't do this, something bad will happen."

"Phycho! Stop being so stupid," says my dad every time I bring up something like this. Remember my dad is a physicist? His faith is in science. He believes there is an explanation for everything that happens, and mere coincidences are not a pattern. He has spent most of his life reasoning things with evidence and research. He never states a conclusion without enough evidence. Obviously, he had to act crazy when I told him that nobody in the school would talk to me if I did not part my hair on the right side.

Initially, he argued that I must feel that way because I have preferred the right side always since childhood. He then wanted to make sure that I approached the problem like a scientist. If I jog my memory, the question was precisely the following. Have you considered **testing the hypothesis** that people avoid you because you sulk the whole day whenever your hair is parted on the left side?

Dad sometimes believes that being unnecessarily rude makes him a cool guy. That is why I decided to ignore him and moved on to the next-best person, mom. By the next best, I mean not that rude. However, she comes with a very generous portion of screaming and an even more quantity of overreaction. Being a psychologist, she follows a humanistic approach. Therefore, she displays a lot of empathy towards her clients. But hey! I am her child. How could I forget that? There was no empathy left for me after the clients.

Nonetheless, there are times when I am considered worthy of a reserved quota if I fulfill the criteria of a clean bedroom, tidy cupboard, and completed homework. Eating vegetables without complaining, cleaning my dishes later, and wearing clothes that she labels decent also do the trick.

"Stop this nonsense, Phycho," she yelled. Yeah, because by her logic, nonsense will not be considered nonsense if you did not raise your volume to scream the word *nonsense*. She then threw a somewhat similar question at me. What evidence I had gathered to support this claim was her question.

One might think the study of psychology is more philosophical, but no, it also thrives on research and evidence.

This time, I was ready with the evidence. I said that my hairstyle was the only thing different about me the first time it had happened. The second time it happened, I had changed my hairstyle after recess. Before this, people laughed at my jokes instead of laughing at me, which became the case later. The third time was recent. I had purposely been on the best conduct coupled with my best smile and the perfect amount of enthusiasm. This was right after I had spoken to dad because his question was challenging enough to give it a shot. Remembering all the details, I explained that I am essentially not a superstitious person. To make her understand me better, I also delivered the famous dialogue from a crime movie. It goes as "Once, a surprise; twice, a coincidence; but thrice, and it is a pattern." Her expressions while I explained kept getting weirder by the second. She then laughed at me. I was perplexed. She then said, "Okay, I will take your word for it but would you please just confront your friends with the issue?" This was tough. I was nervous, imagining how my friends would react to such a question. I knew if they did not want to be my friends because of my hairstyle before, now the audacity to confront them about their judgment would just push them off the edge.

I thought I should rather part my hair the way they liked every day if it meant I still get to have friends. The moment I paid attention to the imaginary conversations I had with my friends in my head, I realized that I sounded a bit funny. Because, c'mon, I know I am still a charming person despite the funny hair. And also, a friend should never judge you by your looks. My friends would have had to just accept me.

Guess what? They both made a fair point. When I finally confronted my friends, the conversation got real awkward real fast. They made fun of me to my face. They told me that I looked horrible no matter which side my hair was parted on and as if that was not enough, one girl added to that, saying she had not even noticed my hair until that day. Everyone, however, later assured me that we were friends with or without hair. It was never about the hair. They showered me with compliments, or let us say, I autocorrected whatever they said to what I wanted to hear. Similarly, the evidence that I had collected was also autocorrected by me but not in a good way. In reality, some were busy, upset, preoccupied with their own problems, or totally clueless about how I felt like I was clueless about their lives. Turns out there is no right side after all. Pun intended!

I learned that **sometimes in life, we are so ignorant of others and unhappy with our own lives because we only choose to focus on ourselves**.

We come up with so many different theories regarding who perceives what. In fact, **what we perceive is also only the tip of the iceberg.** Physics loves theories. Numerous contrasting ideas are put forth by intelligent people to explain a *single* phenomenon. The only one accepted is the one that gives proof or experimental evidence.

A similar technique called **testing evidence** is also used in psychology. The clients are asked to observe, test, and provide some real-life instances as evidence for their thoughts.

Our brain tricks us into arriving at a biased conclusion by relating purely coincidental events because that conclusion seems **more acceptable** to us. It is a defense mechanism. There is also a mental disorder called OCD (obsessive-compulsive disorder), which makes people obsess over a particular thing. It enforces superstitious behavior. **Do you not think we are better off without superstitions**?

Superstitious people are either overthinking something, obsessing over something, or believing in an old, outdated perspective. These superstitions generate out of fear, and then people label certain events as a bad or good omen. There are many examples of the exact same ritual being considered a bad omen in one culture but good in another.

How can that be? That is because people from different regions have different histories of their culture. The stories of our ancestors about the hardships, successes, failures, and experiences are passed down over generations to protect the descendants. But, does everything that worked for them must work for us? No! We have evolved.

Today, we live quite a different life. It is dissimilar to theirs and was probably beyond their imagination. There was a time when physics and psychology were not taken seriously, but (thank God) now they are respected.

I am glad I got rid of my superstition (or, in his words, stupidity). It was exhausting to obsess over my looks because I would end up feeling terrible about myself. And oh boy! Did I love it when I parted my hair on the left side? Yes, definitely! It felt so right. Again, Pun intended. Although now it seems there is one person that hates my hair. When I told her that I was so busy with the research on the hairstyle that I genuinely forgot to turn in my homework, my teacher lost it.

By now, you must have started picturing me in your head as you read. I am sure I look charming in your imagination. Or do I need evidence for that too? Hehe...

Perspective

"See the world with fresh eyes."

The world is full of wonders. Each one of us is unique from birth. The course of our life makes us even more distinct from one another.

Any two individuals can never be the same. A pair of identical twins also have different memories of the same incident based on their individual perspectives, experiences, and behaviors. It does not matter whether they look alike, experience the same events, grow up in the same environment, eat the same food, or do everything together. They will still have different interpretations of life because of how one develops on the inside. We only need to keep our eyes open to spot our differences and acknowledge the similarities in all of us.

Now, if your eyes find it difficult to see this kind of stuff clearly, you certainly need a *fresh pair of eyes*. So, go and shop for them. I mean, get some new fancy glasses. Scientists have invented a variety of lenses because our naked eye has limitations. There are so many things around us that we are unaware of.

My bedroom, for example, appears different to me than it does to my mother due to our different visions. She has a peculiar and naturally enhanced sight that helps her spot dirt even in the cleanest of rooms like my bedroom. I, on the other hand, suffer from a little obstructed vision. I am unable to find anything that I lose in my clean bedroom. Well! I need to update the power of my glasses soon.

Lenses play a pivotal role in physics. **Our naked eye is already good enough, except some things are invisible to us**. How do you study those? Lenses!

In astronomy, telescopes use a particular combination of lenses to observe the movements and changes in planets, stars, and other celestial bodies. Even to look at our Sun, a specialized lens is needed. Let us not forget that microscopes use yet another combination of lenses to observe very tiny specimens like bacteria, viruses, living cells, and many more. Also, our naked eye cannot see all the parts of *the electromagnetic spectrum* like the infrared ray, the ultraviolet ray, or the radio wave. We can only see the visible light spectrum. Just imagine how clueless we would have been about our own surroundings had our physicists not researched optics.

By now, you must have realized how crucial using a lens is. I mean, imagine having a blurry vision while riding a bicycle. Would you feel confident enough to paddle? If you are someone who oozes confidence and decides to ride like that anyway, I wish luck to all the pedestrians on your route that day.

Just imagine the headache a blurry vision causes and the constant squinting of eyes. If you are one of those who avoid wearing glasses because of fashion, I have heard that contact lenses are awesome. These contact lens shops nearby my house lure me in big time. My parents have, let's say, made it pretty clear to me that I am not getting those until I turn twenty-one. Who am I kidding? I don't need them. I rock glasses like a model of Teen Vogue. Now, that's some confidence! (Look what you are missing out on, Teen Vogue.) Imagine the look on the faces of my friends when they see me on the cover. I definitely need to be able to see that look clearly. I wonder how my hair would look on the cover page...the texture of my skin, and color of my skin, my acne, my dry lips...Whoa! I observe a lot (you have seen this in the previous chapter), and I definitely form opinions based on it. Maybe, the *real lens* is in our head.

What on earth is a *real lens*? And why is it stuck in the head? Silly! The lens in the head is our perception of things. We use this lens for judging events, distinguishing the right from the wrong, developing opinions, as well as analyzing what we saw or heard or how we felt. It is our own customized lens.

Psychology calls it the ***subjective reality***. Sometimes we misinterpret, misjudge and make mistakes in processing the information collected from the world around us. This error in thinking is called cognitive bias. These cognitive biases are dangerous and can make us pretty stubborn in some opinions, quite judgmental too.

If we hold **prejudice**, we need fresh eyes. In the world today exist issues like *racism* which is discrimination because of the color of the skin. Also, *sexism*, when discrimination is based on gender. Likewise, *ageism* or discrimination due to age and *linguicism* or discrimination based on how one speaks.

If you are somebody that chooses friends based on the color of their skin, then you need *fresh eyes*.

If you are somebody, who decides which person is right or wrong because you are biased about a particular gender, you need *fresh eyes*.

If you are someone who decides whom to respect based on age, then you need *fresh eyes*.

If you hear someone speak differently and make jokes about them, you definitely need *fresh eyes;* probably, fresh ears too.

Essentially, people today need a new perspective to look at the world. Unfortunately, merely updating a lens on the eyes to see things clearly is not enough anymore. You need to change those *wrong fits* inside the head.

Lenses in your head can be the *right fits* when you judge a person not just by the circumstances but also by morals, ethics, and values, displaying humanity and empathy. Each person is composed of so much more than we can see superficially. Our *right fits* also protect us from threats. It is okay to call out a bully because bullying is wrong. It is necessary to request the court an appropriate punishment for the criminals.

Have you heard someone say "**turn a blind eye**" to something? Mom told me to turn a blind eye to the actions of those identical twins in my class. Tanya hits me. Her twin sister, Sonya, then immediately kisses me on my cheek. This happens every time our teacher turns her back towards us. So now, when the teacher *turns her back on* us, I *turn a blind eye to* the twins.

Dad says it is just how adults turn a blind eye to the news channel when they report a crime or an environmental issue. I feel this is similar to how some adults on the internet have turned a blind eye to wearing any clothes whatsoever.

Dad himself turns a blind eye to an entire human - my mom's mother - by ignoring literally every comment that comes out of her mouth. Mom, too, does the same to his mother. My grannies are super adorable ladies who visit us only during the holidays, cook yummy food and love me a lot. It is why my parents need to fix that blind eye and update all of their lenses *ASAP*.

Are you ready to see the world with *fresh eyes*? We are called the future of tomorrow. *And, the future definitely needs the best lenses ever (externally and internally)*. I believe it also needs Phycho on the cover of Teen Vogue. (Teen Vogue, if you are reading this, know that my dad thinks I look like a cartoon which makes me a superhit television material already.)

Desire

"Be the best at what you do."

Interestingly, once upon a time, this famous statement was used by parents to create ambitious kids. Have you heard this phrase before? Did you feel a certain kind of pressure? Or did you get inspired to be the best? I certainly felt pressured when I heard it for the first time, given that I had just learned the number of the total population of our planet.

Imagine the hard work required in being the ultimate at a skill. How can every person skilled at something be the best at it? Somebody has to be the second for someone to be the first. So, isn't 'BEST' just a relative term and not a superlative word when time is considered? And then...what if I really want to be just 'OKAY' and not the 'BEST'?

Recently, I asked my dad if he was *the best* at physics, and he had this inexplicable, kind of confused expression on his face. My mom heard this question from the kitchen and replied, "Who cares about physics when he is certainly the best at not doing chores and not fixing broken things?" I wondered if she was the best psychologist, but obviously, she wasn't. I could literally see the compassion and empathy getting lost in her duly crafted taunts for her husband. That same husband preaches physics is interesting because it is everywhere around us but doesn't take any interest in the physics that is lying around in our house. For example, fixing the drain or changing the bulb.

Should we even work hard if there is nothing like *the best*? Why not just do nothing and relax? YOLO! People nowadays toss YOLO on social media statuses like free Halloween candies. If you haven't encountered this popular acronym yet, let me tell you *YOLO* simply means, *You Only Live Once*. It finds multiple applications like setting priorities, maintaining enthusiasm, and providing motivation. Although, its primary use is to serve as an excuse for avoiding the tasks that you find boring. I believe the issue is not the hard work. It is actually about how you perceive *today* and *tomorrow*.

Let us rephrase and define our problem correctly. The problem is: When you know life is short, then while making big decisions in your life regarding your career, a relationship, or your priorities, what should you really focus on – a) planning ahead and working meticulously to be better prepared for tomorrow or b) enjoy and live today because tomorrow is very uncertain?

Say, traveling to your dream place is one of your life goals. To afford the trip, let us assume that you will need a lot of money. For that, you will have to struggle day and night to gather enough savings. At the same time, you wish to enjoy your life daily by working less, giving time to your loved ones, and enjoying your leisure time. We all know that the present is what we truly have. Should you even plan for something uncertain like the future? But then, should you never aim for the dream goals because of the sacrifices involved? It is, indeed, a difficult choice because today is the *gift of time* but tomorrow withholds the *magical possibilities*. However, **if life is short, it will not take that long for tomorrow to become today.**

What procedure do physicists follow to gather evidence? First, they define the problem. Then, they develop the objectives or goals of the research. They then design an experiment (planning). Next, they gather data by performing the experiment. Finally, they interpret the data to state their conclusion (about whether they actually found what they were looking for.) Scientists toil day and night, face multiple failed attempts, learn from their mistakes and keep modifying plans.

Physics teaches us **planning, learning, and working today for that bright tomorrow**. We all have so many things that we want to do for ourselves and others which is only possible if you hustle today focusing on the big picture.

Psychology, too, requires a lot of research, and it is done in the same way as in physics. The counseling part in psychology is different. It relies on instincts and interpreting the non-verbal cues, the hidden layers of feelings, the body language, and the mixed emotions. In counseling, the therapist focuses on helping clients develop an awareness of themselves and their surroundings. They study the influence of people around them, the meaning of situational behavior, and so on. For well-being, a crucial realization that needs to be achieved is how much emotional value we give to things in life and how much is actually appropriate. An approach in psychology called the Gestalt approach follows a **here and now** concept. It emphasizes living in and experiencing the present moment. It also focuses on the holistic growth of the client, that is, growth in all aspects of life. Living today to the fullest is, thus, essential for proper functioning.

What's our conclusion? Physics teaches the importance of hard work today, but Psychology teaches living wholly in today. **It is hence not a choice between the two, but the *order***:

1) **Plan and prepare for that dream**: Plan *smartly*. Set *manageable tasks* and *realistic plans* with some time reserved for leisure too. Then, prepare! If you are not ready for tomorrow, the moment an obstacle arises, you will feel that you should have thought about it in advance. When rigid plans start disappointing you, slow down and modify. That is evolution! If our ancestors had not revised their plans, we would probably still be running naked in jungles, hunting for daily food. There will be some sacrifices on the way but remember the big picture. Our smartphones today are also a result of dreams, planning, and hard work.

2) **Now detach from that dream**: Planning never hurt, but failing in those plans hurts, right? (After you are done planning, dedicating your time, and utilizing your energy) Stop getting emotionally dependent on that outcome of tomorrow as you cannot dictate your

future nor control everything. Remember, the particular result that you desperately await doesn't define you. Start living your today, value your life and make amends for those sacrifices you have had to make while working hard in the remaining time available to you. Who said working hard for your dreams means not living your life? Live the work. Enjoy the journey!

3) **Ultimately, after every big action/step in life, say,** *"If it happens, I am okay; If it doesn't, I am okay."* Physical and mental work finds its balance in emotional well-being. Detachment from the outcome gets you that well-being.

Today, mom surprised me with two things: first, a pizza, and then, a wash-all-the-dishes-later protocol. Don't worry, like dad, I may not be the best at chores, but I came up with the best pocket money plan ever. I am already working hard, and as per the plan, soon I will have enough to purchase a dishwasher. Beat that, mom! Meanwhile, "I am OKAY!"

Identity

"What will people say?"

Do you ever wonder how much free advice one individual can offer? My neighbors, relatives, friends, parents, dentist, the watchman of our building, and even his wife have so much free advice on improving my life. So, basically, everyone I know loves to shower me with advice from time to time. These pieces of advice concern which exact career I must pick, which person is worth dating for me, and even how many baths I need to take to look hygienic.

Imagine how I must feel when their voices shouting "Aim higher, aim higher!" play in my head like a background score every time I decide about my career, a friend, or even a simple bar of soap.

My mother was reading this as I was typing, and she looked more freaked out than I was. Then, I was summoned in the kitchen by #eyebrowraised #scarystare face. The utensils she had been washing, all of a sudden, started looking so shiny that the bounced off light felt blinding. I could not predict whether my writing had given her some much-need energy or unnecessary frustration.

Without realizing it, I had entered a discussion and begun washing dishes alongside her. That is when she described a very amusing theory of psychology called the ***self-actualization theory***. Self-actualization is the tendency to become your best version.

Humans, according to my psychologist mother, strive to do better. This need keeps them going, changing and evolving. It is healthy to aspire and set goals. Here is the tricky part, though. What happens when you fail to achieve the goals you set for yourself?

She explained that there are two different versions of how we perceive ourselves. One is our ideal self, which is how we want ourselves to be; the other is our true self, which is who we actually are. The more the gap between the two versions, the more we get upset with ourselves and our lives, believing we are not enough or not doing enough.

I decided to consider my teacher, Ms. Ruby, as an example. She is just exceptional at teaching. I always understand every single concept in her class. At the same time, she is also famous, or rather infamous, one may say, for being excessively strict. It makes me believe that she thinks her ideal self includes being the meanest, most ruthless, and the strictest teacher to ever exist.

Well! I would like to tell you that if what I believe is true, the gap between her ideal self and her true self is non-existing. However, recently, I witnessed her hesitantly showing compassion towards a student when he cried in front of her. She found herself forced to offer him a tissue. Maybe it finally dawned upon her that she teaches a generation of humans and not robots. That day must have been so difficult for her, having let herself down. She had, at last, presented the sensitive side of herself although unwillingly and accidentally.

I then dared to consider myself too as an example. I want to be a pop singer. My aunt says while deciding career, you must always aim higher than everyone else because only then you will be successful and happy. I wonder if following the footsteps of Michael Jackson or Lata Mangeshkar would be it. I mean, how much higher than that can you possibly aim for in singing? After conducting a lot of research, it turns out that their lives were not so easy and tempting after all.

Honestly, I would not want to live a life of a copycat. If a path has worked for someone, it does not have to work for me. I want it my way, making mistakes and not comparing my aiming potential. Learning about these two people taught me that, throughout their careers, they were so much more than just singers. They happen to be famous for that one skill, but, as a person, they have so much credit. As a whole, they are known for their work ethics, willpower, principles, and their contribution to society. I feel they became the best versions of themselves. That makes me want to be the best version of myself, whatever that is, by setting my preferred level of goals.

If one must aim higher, it should not be only for the career. In a true sense, we must realize what *unique value* we can bring to the world and humanity.

According to psychology, we are already hardwired to do so. However, we must be cautious about how unrealistic our expectations of ourselves can be. We will not always get everything that we want. In the previous chapter, we learned that being the best depends on the reference frame. Nonetheless, being the best version of oneself is definitely within our reach.

We should all strive for that in the utmost capacity. Right? Although, I feel I will be letting my aunt down when I fail to fit *the idea of what I should be* that she has created in her head. To that, my dad says, "What people think is useless." He uses a phenomenon in physics as an analogy to support this claim: The Doppler effect!

Doppler effect or Doppler shift is the apparent change in the frequency of light or sound when the object moves closer to or farther away from you. His favorite example is the stars. The stars that appear red in color are actually moving away from the earth (us). The ones that appear blue/violet are approaching us.

In summary, the light we perceive is irrespective of the actual color of light emitted from the star, depending on whether it is approaching or receding from us. The farther a star moves, the frequency of the emitted light decreases. This happens because as the space increases between the star and us, the emitted rays become less crowded or less frequent. The frequency of the red color in the visible spectrum of light is the lowest. (Next time you spot a red star, do not forget to say *Bye-bye*. A blue star would mean *Hello there.*)

Now, my dad says blue is a nice color. So, whenever I, like the star, approach somebody's *ideal picture* of me, they feel something right or acceptable is happening. Contrastingly, when my actions differ from their expectations of me, it is as if a red-colored warning sign flashes for them. My aunt must be seeing those warning signs every time I talk about my career. So, when she feels I am not aiming high enough, I am merely exploring other aspects of my life at that moment. Dad says, I, the star, remain the same person no matter what others say about me.

"Phycho, right now, you must be liking a few things about yourself (blue) and hate a few (red), but you still remain the star that you are underneath it all. And, you will always glow."

This happens to be one of the kindest things he has ever said to me. The other day, he said, "You are a good child." That was for fetching the TV remote while he rested lazily on the sofa.

Remember, you will continue to glow brighter day by day, in *different things*, in *different ways*, the more you *start exploring life*.

My parents, in fact, had a debate later. Mom exerted that one should focus on why those words matter to us at that moment instead of focusing on what those words were. Dad said, "People have mouths. They will use those. What we can do is decide when to use our ears." Mom then remembered how this week Dad was always wearing his earphones. Moments later, he was found washing the dishes.

Death

"Have some faith!"

Do you have faith in something? It may be faith in yourself, a religion, parents, friends, work, or even your pet. I do. I have faith in food. I am not trying to be funny; That is just how it is. I believe that the right food can lift your mood, change your perception of the situation, retrieve enthusiasm, and make your brain work in miraculous ways that it already does by coming up with the solutions to the problem that you are stuck on.

So, whenever I am stuck with my homework, mom asks me to rest (sleep) or have food. The latter always works because the former is nowadays kind of my routine and hence, lost its magic.

What is faith? Faith is a strong belief that gives you hope even when things don't make sense anymore, even when you don't see a direction, and even when everything is going wrong. As a physicist, dad's faith lies in science. But whenever his faith in science is shaken, he also has faith in the supreme power, God. He feels that it brings him sanity.

My mom, on the other hand, has faith in herself. She believes we have all the answers that we seek within ourselves, and if we calm ourselves and believe in ourselves, we start noticing those. Being a therapist, she obviously believes in the innate capacity of a human being. She, too, worships God, but that faith is heavily shaken whenever my teachers report my mischievous behavior because she realizes that even God can't help her in my case; only she can. Evidently, that's true because fear of punishment by God is nothing compared to the fear of getting punished by an angry mom.

Everybody has a different faith, but this faith is challenged when you face a real-life crisis like losing a person, witnessing your pet's heart stop beating, or running out of money during an emergency. The death of someone you love is painful. Getting over the pain is extremely difficult. I cried and cried when my friend's grandpa passed away, although I hardly knew him. I was only seven years old then. I had started fearing that my grandpa, who was older than him, was next on the list and maybe just got lucky enough to miss his turn. I was terrified of losing him, and seeing my friend in immense pain was painful. I remember everything about this experience so distinctly because I had started having nightmares. I would call grandpa every night until he stopped receiving my calls. He was annoyed and out of creative answers to my constant questions like, "Grandpa, how long do you think you will take to die?" Now I realize how that question would not have been so correct. It would make my mother so angry. She didn't want to lose her father, but she was worried about me more because I just was too stubborn and insisted on doing whatever felt right to me. To me, it felt like I was saving him in some way by confirming his existence with my calls, and if I stopped, he might just go away. How was I supposed to live with the regret of being sloppy? My dad, then, one day, came up to me to explain what death meant. It was, however, in terms of science using his knowledge of physics.

Physics says everyone and everything is made up of energy. We - the living things - have energy, and so do the non-living things. Energy can be obtained and transferred. When a ball is raised to the top of an inclined plane, it stores **potential energy**. As soon as it is set in motion, that energy is converted into **kinetic energy**. Molecules need sufficient energy, known as **activation energy**, for their job, which is to break bonds when they collide. We, too, need enough energy for our daily life functions.

Dad says life is also a form of energy. He calls it *living energy*. It is generated and used when we eat and digest food, exercise, or simply breathe. Whenever I cry, he asks me if I feel exhausted, and of course, I do. He then says, "That's because you have spent plenty of energy. You need to have food to restore that lost energy." That is how I got my faith in food. It gives me my *living energy* and helps me stay healthy.

Psychology, too, gives us a related theory. Freud, a remarkable psychologist, states that there are two basic classes of instincts: **Eros, the *life instinct*, and Thanatos, the *death instinct*.** The *life instinct* pushes us towards survival. The *death instinct* makes us feel depressed and hopeless.

Mom always says one also needs to maintain a healthy mind alongside the body. She tells me to practice good thoughts, read inspirational stories, feel motivated about work, and always be kind to others. She made me experience how paying attention to others and caring about them makes a huge difference in my own life as it does in theirs. As decided, every year, we donate my old clothes and gift some new ones to less fortunate kids. I certainly feel a lot better, less depressed, more grateful, and utterly satisfied. However, how is one supposed to feel better when someone dies? What helps with that kind of pain?

Psychology tells us that people go through five **stages when dealing with grief**: 1. *Denial*, 2. *Anger*, 3. *Bargaining*, 4. *Depression* and 5. *Acceptance/Hope*. After the last stage, people have officially moved on with their lives. They, then, start having faith in something new they want to believe in and embrace life. Mom said humans have also developed many *defense mechanisms* to face the harsh truths of life. One of which is *rationalizing* or coming up with a reason for the occurrence of something. Another one is *denial* or delaying the acceptance of the fact. Yet another is *sublimation* or expressing the emotion in an acceptable manner. For example, playing sports to distract yourself from the anger.

Now that Physics bluntly made us understand what supposedly death is and **Psychology helped us understand stages of the process, what is the solution to end such grief? Psychology told us that the *key* is acceptance.** We *have to learn* to accept. The sooner you *accept*, the sooner you heal. With acceptance come hope and faith. But how do we speed this process? Nobody wants to stay in grief that long. Is there something to speed up the process of gaining back that faith? Yes! You must start by telling yourself to have some faith, even if you believe it is impossible for you. You must have had people say that to you when you were in doubt or pain. In this case, people are right.

The only way to end grief is to, slowly or quickly, start accepting what is in front of you while telling yourself to have faith until you really do.

Exactly what amount of food helped me get over my friend's grandpa's death or my fear? Zero! I completely lost my faith in food after having eaten a lot. However, as I kept reminding myself to have faith, I discovered faith in something else. I *chose to believe* my teary-eyed dad when he said, "Everything will be alright." My trust then lay in my parents. Eventually, time did its magic. It healed the pain. Meanwhile, I also believed in my ability to get over the pain. So chronologically, I had faith in food, my parents, myself, and many other things until it shifted back to food again. This time a little healthy food.

Understand that, in life, knowledge will help you, as will experience. However, nothing will teach you to escape problems. You still need to face things and hang on. How else would you evolve? Then again, you never know who or what might just come through for you and when. For you, I am trying.

Epilogue

Dear Phychos,

Our journey has now come to an end. You have learned some new stuff but know it is difficult to embody. With time, you will. We started discussing life and progressed to discussing death. I got no idea about the afterlife concept. So, I won't go there, but if it did exist, it would be interesting to someday test if physics and psychology were to apply to the problems there too. We all have so much to learn. So far, I have learned that the true purpose of life is to keep living, experiencing, and learning. You can, then, choose to share that knowledge if you like.

In the end, we all are one and the same. Life has so much to offer to us – the good and the bad. When things go wrong, you may hear some say that everything has a reason; everything has its own timing; everything falls into place in the end; some things are just not meant to be... Others say nothing. These people just keep living their lives without ever giving problems a second thought. Not everyone can afford to ponder over the instances in their lives. We all have come across such people or been those people. We get so absorbed in a busy routine that there is no time left to consider our emotions and feelings. It is indeed a gift if you receive the time and the ability to understand an experience. One only comprehends how necessary and liberating the whole emotional process was, after completely living it; having had experienced what it is like to dwell in grief, to enjoy an achievement, to foster curiosity, to engage in some meaningful conversations, and also to carry a heck lot of confusion in mind. Every moment, interaction, experience, and lesson that is yours should be worthy of your appropriate attention.

Each life has been and will always be different. However, from a macroscopic view, it is one and the same. We all take birth, and we all die. The start and the end are the same for all, but the middle makes all the difference. We all have been given human life. We happen to be the best-evolved species. So, why not make the most of it and even set a tone with our compassion, intelligence, and complexity? What we do with our lives is up to us. It is up to every individual to make their own choices and direct their own course. You don't owe your life to anyone but yourself. So, love yourself, forgive yourself, improve yourself, test yourself and trust yourself. It is only then that you can, in a true sense, help others do the same. These are the real goals that our generation needs to emphasize. Unlike we might be told, earning a lot of wealth, gaining recognition, or constant mindless hard work are not real goals that make you truly special. It is up to us to redefine them. Holistic growth as a person, satisfaction in yourself, and sharing of your wisdom with others - those around you that make you who you are, add value to your life, and exist just like you do - is what makes the human process of living a miraculous gift. Some of us might be lost in our ways due to the chaos of life. With time, patience and hope, we need to push hard to get in touch with the core of our existence. Remember, we are all in this together. So, be kinder. Also, be wiser, braver, better, fitter, and most importantly, your own self.

I can just keep adding more cool philosophical phrases, but I stop. I guess if you already have decided to grasp something from this journal, you don't need those phrases. You only need deep breaths. No more philosophy then... or psychology... or physics. For you, only lots of love, hugs, and smiles!

Acknowledgements

From an idea to a book, it is an experience I cannot justify with words no matter how much I have tried. It is my own to cherish, and I am immensely grateful for it. However, all of it was made possible by certain people who kept me afloat whenever I lost my balance on this track. My first thank you is to my partner. Thank you, Pushpendu! Your friendship means a great deal to me. In troubled times, it was your voice that finally made me jot down my unbearably uncontrollable thoughts. Ideas from which later evolved into *Phycho*. You joined me on this journey as an illustrator but ended up taking a lot more roles until the finish line. I am thankful for my sister, and primary editor, Asmani, who read each chapter the minute it got typed. Without her harsh but frank comments, I would not have taken this skill seriously. Although this book is a surprise for my parents, anything I would ever do is impossible without the wisdom (in form of nagging, mostly) they impart on me. My courage to take risks is because of my huge family that has always supported me in whatever way possible. A big thank you!

The next batch of acknowledgements is for a miscellaneous group of people around the globe. I will always be grateful for my grandfather, Ashok, who slept while reading the first draft and justified the act by telling me the draft was boring. Had he not been so honest, publishing would have stayed a dream. Seeya, my Australian sister, whose love for reading is similar to our grandfather, became my proofreader. My wide friend circle spread over different cities, though I kept most of you clueless about the book, I owe a lot of things to even your mere presence in my life. My teachers and mentors, your sincere efforts have developed me as a person. Thank you all! Ms. Wieben thank you for cultivating my aptitude in science. Mr. Kanitkar, Mr. Meyer, Mr. Ruffle and Mrs. Nagesh thank you for noticing and appreciating my writing. Mr. Sarolkar and Mr. Dhage, thank you for guiding me in my case studies.

***

The journey from the birth of an idea to a published book has a small group of people working tirelessly for a limited period. The journey from a published book to the heights it can achieve has a growing group of people displaying gracious effort for an indefinite period. Something from *Phycho* that I always found difficult was putting faith in the miraculous power of time. Therefore, envisioning a possibly realistic future, I would like to thank each one of you in advance who will now join me in this next journey. Lastly, my deepest gratitude lies to all the children for whom my love would never cease. *You* inspire me, provide hope and drive me to create a better world for you.

www.ingramcontent.com/pod-product-compliance
Lightning Source LLC
Chambersburg PA
CBHW031444210526
45464CB00005B/2328